My World of Science

FORCES AND MOTION

j531
.6
R816f

Angela Royston

DISCARD

Heinemann Library
Chicago, Illinois

WILDER BRANCH LIBRARY
7140 E. SEVEN MILE RD.
DETROIT, MI 48234

10/03
WX

ublishing

nal Publishing,
Chicago, Illinois

Customer Service 888-454-2279

Visit our website at www.heinemannlibrary.com

All rights reserved. No part of this publication may be reproduced or transmitted in any form or by any means, electronic or mechanical, including photocopying, recording, taping, or any information storage and retrieval system, without permission in writing from the publisher.

Designed by bigtop
Originated by Ambassador Litho
Printed and bound in Hong Kong/China

06 05 04 03 02
10 9 8 7 6 5 4 3 2 1

Library of Congress Cataloging-in-Publication Data
Royston, Angela.
 Forces and motion / Angela Royston.
   p. cm. -- (My world of science)
Includes bibliographical references and index.
  ISBN 1-58810-240-8 (lib. bdg.)   ISBN 1-4034-0039-3 (pbk. bdg.)
 1. Force and energy--Juvenile literature. 2. Motion--Juvenile
literature. [1. Force and energy. 2. Motion.] I. Title.
 QC73.4 .R69 2001
 531'.6--dc21

Acknowledgments
The author and publishers are grateful to the following for permission to reproduce copyright material:

P. Aikman/Trip, p. 27; Trevor Clifford, pp. 14, 15, 17, 18, 21, 22, 23, 24, 25; Corbis, pp. 8, 11, 16, 19, 28, 29; Robert Harding, p. 5; Maximillian Stock, p. 7; Powerstock Zefa, p. 10; Dr. Marley Read/Science Photo Library, p. 6; Rex, p. 20; H. Rogers, p. 4; Frans Rombout/D. 0. Bubbles, p. 26; Stone, pp. 8, 12, 13.

Cover photograph reproduced with permission of R. Smith.

Every effort has been made to contact copyright holders of any material reproduced in this book. Any omissions will be rectified in subsequent printings if notice is given to the publisher.

Some words are shown in bold, **like this.** You can find out what they mean by looking in the glossary.

Contents

What Is a Force?

A force makes things move. These people are moving a piano. One man is pushing it. The other man is pulling it.

Pulls and pushes are forces. This girl is pushing down on the pedals to make the bicycle move forward.

Machines and Forces

This bulldozer is pushing earth and trees out of the way. Machines have **engines** that make the force to move heavy loads.

This crane is lifting a heavy load. The engine makes a force that moves the jib. The jib pulls up the cable. The cable lifts the load.

jib

cable

Natural Forces

Wind and moving water are powerful **natural** forces. Wind is air that is moving. It can bend trees and push leaves through the air.

Moving water also pushes
things. This boat is
floating down the river.
The moving water is a
force that pushes the boat.

Moving Your Body

Our **muscles** make our bodies move. This climber's muscles **contract** to make force to pull her up the **steep** rock.

You can move in many different ways.
This woman is swimming. The muscles
in her arms, legs, and feet make a force
that moves her through the water.

Stopping

Forces can also be used to make something stop moving. This dog wants to move forward. Its owner is pulling it backward.

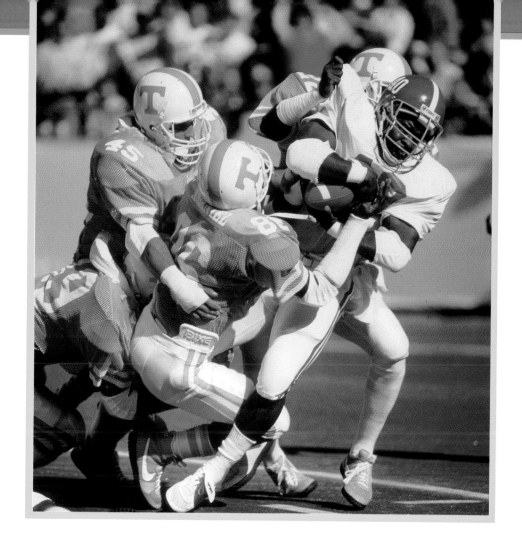

Pushing or pulling something that is moving can slow it down or stop it. The players in yellow are pulling the player in white. They are trying to stop him.

Changing Shape

Forces can be used to make some things change shape. Soft clay is easy to push and pull into many different shapes.

This boy is squeezing the empty carton to push the air out. He is making the carton flatter and smaller. Now it will take up less space in the trash can.

Changing Direction

Forces can make something change direction. This tennis player pushes his racket against the ball. The ball changes direction and goes back across the net.

Forces can also make something turn
in a circle. You have to twist the top of
a jar one way to take it off. You twist
the other way to put the top on.

Getting Faster

The harder you push something, the faster it moves. This girl is pushing a toy train across the floor. If she gives it a bigger push, it will move faster.

These runners are working hard to run
as fast as they can. Their feet push
down and backward on the ground.
This pushes them up and forward.

Slopes

A slope can change how fast something moves. This girl is skateboarding down a slope. The **steeper** the slope, the faster she will move.

This girl is pushing her wheelbarrow up a slope. She will have to push harder than she did on flat ground.

Friction

Friction is a force that slows things down. The boy pushes the toy and then lets go. The toy moves quickly at first. Then it slows down and stops.

The toy slows down because the wheels rub against the ground. The rubbing is called friction. The wheels on the toy are **rough.** They cause more friction than smooth wheels.

Testing Friction

This boy is using balls and a **ramp** to test which kind of floor has the most **friction**—carpet or wood. He measures how far the ball rolls.

He finds that it rolls farthest on wood. The **rough** carpet causes friction. The wood is smooth. So there is less friction between the wood and the ball.

Using Friction

You can use **friction** to slow yourself down on a slide. Push your hands and feet against the sides of the slide. The friction will slow you down.

Bicycle brakes use friction to make the wheels stop spinning. When you pull the brake handle, two rubber blocks rub against the wheel.

More and Less Friction

The **soles** of your shoes are **rough.**
They cause more **friction** between your
feet and the ground. The friction keeps
you from slipping when you walk.

The less friction there is, the more you slide. Snow is very smooth and creates very little friction. Skiers slide fast across the slippery snow.

Glossary

contract to become smaller

engine something that uses electricity or fuel, such as gasoline, to make a machine move

friction rubbing between one object and another that slows movement down

muscle part of your body that helps you move

natural something made by nature, not machines

ramp slope made by people

rough bumpy or uneven

sole bottom part of a shoe

steep when a slope rises or falls very sharply

More Books to Read

Gibson, Gary. *Pushing and Pulling*. Brookfield, Conn.: Millbrook Press, 1996.

Hewitt, Sally. *Full of Energy*. Danbury, Conn.: Children's Press, 1998.

Marshall, John. *Go and Stop*. Vero Beach, Fla.: Rourke Book Co., Inc., 1995.

Index